YOUR KNOWLEDGE HAS VALUE

Bibliographic information published by the German National Library:

The German National Library lists this publication in the National Bibliography; detailed bibliographic data are available on the Internet at http://dnb.dnb.de .

Imprint:

Copyright © 2015 GRIN Verlag
Print and binding: Books on Demand GmbH, Norderstedt Germany
ISBN: 9783346155320

This book at GRIN:

https://www.grin.com/document/542856

Klaus Spicher

The Heuristic Business Forecasting Methods Revinda and Metrix

GRIN Verlag

GRIN - Your knowledge has value

Since its foundation in 1998, GRIN has specialized in publishing academic texts by students, college teachers and other academics as e-book and printed book. The website www.grin.com is an ideal platform for presenting term papers, final papers, scientific essays, dissertations and specialist books.

Visit us on the internet:

http://www.grin.com/

http://www.facebook.com/grincom

http://www.twitter.com/grin_com

New Forecasting Approaches

Cybernetic Model Approach
&
Similarity Metric-based Method

by

Prof. em. Dr. Klaus Spicher

Summary / Intention of the Paper

The report describes two new heuristic approaches to time series analysis and forecasting for business purposes. Both approaches avoid any assumptions according to assumed process attributes behind the data (e.g. stochastic process, stationarity, normal distribution of random noise, ...). Those methods just engineer data of any kind of business processes. – Only unidentified (inherent) process structures are used for forecasting.

Key words: Revinda, Metrix, Business Forecasting, Similarity, Model-free forecasting Methods

Table of contents

1   Introduction... 3

2   Brief Forecast-Methodology Overview 3

3   Business Forecasting 4

4   Comments on Forecast Accuracy 5

5   New Methods supporting Business Forecasting 6

   5.1   Business Process Forecasting - The REVINDA-MM-Approach.................................... 6

   5.2   "ERP-Type"- Forecasting   -   METRIX Approach 8

6   Conclusions:... 9

7   Literature:.. 10

Appendix.. 11

1   Introduction

Speed represents the driver of current business development. IT is about allowing for automatic self-synchronizing (production) processes. Big Data potentially offers the identification of hidden structures enabling for further process improvements. In many companies mobile information access is being used. Multi-Channel B2C, B2B and M2M are gaining the managerial pole position. - But nevertheless the quality of data is the key for producing excellent results. We cannot deny that GIGO determines the outcome. So, it is important that planning is based on as 'realistic' data as possible. After roughly more than 35 years Business Forecasting is back in the focus.

According to [2] "Industrie 4.0" represents the successor of the CIM-Conception of the late 1970-ties, now based on advanced IT infrastructures. In history of Business Trends the predecessor of CIM was the introduction of "quantitative data" for improving business processses. At that time many extrapolating forecasting methods have been developed (Holt/Winter, Lewandowski, etc...), which up to now are standard in estimating future data of business processes. The Helix 1 and Helix 2 cycles [2] describe the development of business trends. The Helix findings can be interpreted as a structural forecasting approach.

2   Brief Forecast-Methodology Overview

Forecasting represents a challenging task. Each problem is linked to special requirements, restrictions, available data and the expertise of selecting the most appropriate method. Common to all is the need for "good results". There is a great number of forecasting methods available. Some of them are standard for certain categories of forecasting tasks. E.g. most ERP-Systems offer 10 or more (simple) standard forecasting methods for routine planning operations. (In this report this kind of forecasting will be called - in line with present penetration of ERP's - "ERP-Type" forecasting.)

The list below specifies a number of well known forecasting methods:

- Simple Extrapolations (e.g. EXCEL-Functions for Trends)
- Decompositions of Time Series (Level, Trend, Season, Random Noise)
- Smoothing Methods <Filters> like Moving Average (MA), Exponential Smoothing, Holt / Winter extension for Trend and Seasonality
- Box-Jenkins Method and advanced auto-correlated ARMA and ARIMA-Models and ARCH, GARCH
- All kinds of Regression Analyses (Least Square Estimators)
- Econometric Modelling (Kalman-Filter, Markoff-Chains, ...)
- Neuronal Networks (MLP, SOM, ...)
- Croston-Method for intermittent/sporadic demand
- Diffusion Models (Bass-Models) for lifetime analyses
- Black-Scholes Equation for financial analyses
- Evolutionary Approaches (Delphi-Method, Scenario Planning)

The list – of course – is not complete. Different software packages (e.g. "R", "MATLAB", …) offer sets of standard forecasting approaches and /-models.

3 Business Forecasting

This paper covers forecasting approaches dedicated for improving planning quality. As planning is the backbone of operational business success, forecast quality is put in focus. Roughly spoken there are 2 main categories of forecast scenarios now being discussed:

In this paper Business Forecasting distinguishes 2 sub sets:

        a) Business Process Forecasting (MM-Forecast) - see [1]
        b) "ERP-Type" - Forecasting

a) Forecast as a Business Process (MM-Forecast[1])

To give a B2C example: A LSP[2] requires from his customer (a B2C company) the forecast of daily picks for two weeks ahead for capacity planning, because the 24 hours service only is guaranteed in case the forecast meets the range between 75% and 120% around actual orders. – Of course the B2C-company knows the order pattern from past experience. But daily B2C orders are influenced by standard and special promotions of Marketing. Another aspect affecting daily sales is given by temporary top selling items and/or the launch of new products, etc.…. The effects of promotions and special market conditions are estimated 'manually' by Marketing.

So, the final forecast has 2 input sources: a "basic process forecast" and the impact of special company activities. Under those scenario conditions a manager will be responsible for the final forecast quality and data maintenance. So, the final forecast summarizes the basic forecast plus – in this example – Marketing "brain-input". – What about the data to be used in next year's forecasts? As top selling items and special promotion will not occur next year at the same period the actual data have to be re-adjusted for providing high quality input data for next year's basic process forecasts. Therefore it makes sense documenting the promotional and special market effects in a log-file. In case of daily forecasts the re-adjustment of weekly data is sufficient.

The same organization holds in case the company also runs a retail chain – e.g. with some hundred outlets – to be delivered on a daily basis as well (B2B). In this case additional factors also will have to be considered, e.g. the volume of allocations fixed by the central purchasing department which again represent part of the "brain-input" to the final forecast.

---

[1] **Man-Machine Forecast**
[2] Logistics Service Provider

b) "ERP-Type"-Forecasting

The second scenario represents the procedure of applying standard forecast methods be-ing provided e.g. by ERP-Systems. Standard ERP-Systems offer a list of 10 or more diffe-rent standard forecasting methods. During the first implementation of the ERP-System the most appropriate forecasting method(s) are carefully selected for going live. But those forecasting methods mostly remain active even the business conditions have changed after some time. The complexity of ERP-systems often represents the key obsta-cle for re-adjusting the selection of "best" forecasting methods in time. So, the quality of forecast data is suffering and thus the complete planning process.

The above scenarios usually are applied on calendar-periodical processes, analyzing time series according to decomposition approaches for Level, Trend, Seasonality and Random noise. (Some ERP-Systems also offer forecasting methods [like the Croston-Method] for intermittent / spo-radic demand).

4 Comments on Forecast Accuracy

This paper is dealing with short term forecast accuracy only, i.e. 1-3 periods (months/weeks) ahead. Daily forecasts are calculated as break-downs from weekly forecasts – based on daily patterns which might change over time within the calendar periodicity.

The standard measures of forecast errors (MSE, MAD, MAPE and TS) will not be discussed here. For the methods we are going to present in this paper MAPE (t_i) will be used and the Tracking Signal (TS) will be compared with a new "Structural Tracking Band" based on historical data).

All forecast approaches to be presented in this paper are based on data attributes and patterns identified from existing historical data only. No modeling is applied. For identifying different categories of attributes a special tool-kit has been developed including relevant statistical tests. Let's take two extreme examples: (1) Historical data represents a (periodic) function; e.g. $f(x) = A*sin(a*x +b)$. This implies that there is no random noise at all $[\varepsilon = 0]$. Do we expect that the forecast approach can identify the structure reacting with forecasting the same function (MAPE$(t_n) = 0$)? In the other example (2) historical data represent a random process $[\varepsilon = 100\%]$ with for instance $X_t \in \mathbb{R}[A; B] \forall t$ - or any other distribution. What about the structure of the resulting forecast data and MAPE(t_n)? Answers will be given as far as available when discussing the methods in detail according to ongoing research.

The problem behind is the question: Is it possible – and IF, how – to forecast forecasting accu-racy for Business Processes, given *in-sample* historical data sets without specified (stochastic or other) pre-conditions? From examples it can be seen, that correlation – especially in calendar-periodical *in-sample* data enabling comparison of two past data periods – represents a rough determiner for complexity and predictability. J. Garland, R. James and E. Bradley [3] present an overview about model-free quantification of time series predictability based on entropy. They show that predictability might not be fully exploited by the forecasting method applied from modeled relation between WPE (Weighted Permutation Entropy) and MASE (Mean Absolute

Scaled Error). For more details using entropy as measure of complexity of time series see [4], [5] and [6].

Periodical surveys (e.g. M-1/-2/-3-Competitions in 1982, 1993 and 2002) - see [7], [8], [9], [10] – compare the accuracy of forecasting methods being applied on a variety of data sets. Those surveys highlight specific forecasting problems and (most) appropriate model/method selection. Interesting comments on the results of M-Competitions related to business forecasting are given by S. Kolossa in [11].

The purpose of this paper is contributing to solutions of practical (business) forecasting problems by providing simple heuristic approaches (aside of scientific approaches) following the practitioners' principle: "Nearly right beats exactly wrong!"

McCarthy et al. [12] show that in spite of increasing computational power forecast accuracy seems to be deteriorating. Unfortunately no comparison between 2006 and now has been found. The reasons for deterioration seem to be globalization of markets, increasing complexity and speed of business processes, decreasing customer-/ and brand loyalty, shorter product life cycles and others more. On the other hand other surveys show that for selecting forecast tools forecast accuracy remained the priority 1 as well for managers as for practitioners [11] over the last 2 decades.

5   New Methods supporting Business Forecasting

5.1   Business Process Forecasting - The REVINDA-MM-Approach
The name of the approach represents a short cut of "**Rev**erse **In**dex **Da**ta Transcription".

Forecast options:
REVINDA has been designed for short-term forecasts, i.e. up to 3 periods ahead. This means 3 months or 3 weeks ahead. The weekly forecast can be fractionized to daily forecasts using daily patterns. Calculation of daily patterns will not be presented here due to complexity of calendar synchronization (moving holidays, etc...).
In addition, the design of the REVINDA approach allows for providing the set of forecast values for the complete forecast-period H_0 on top of 1 -3 periods ahead.

Data requirements:
Two basic periods of calendar-periodic data are required. Business forecasting allows to restrict the periodicity to subsets of data (e.g. monthly data for 2 years, weekly or daily data accordingly). In case of intended daily forecasts, the year will be represented by 13*4 weeks for calendar synchronization reasons.

Notations:
1   n, t describe the basic period length (e.g. n=12 or 13 for years; n=52 for weeks; and 365/ 366 for days), while {t=1, 2, ..., n} represents the time axis.
2   Historical periods are denoted by H_{-2} and H_{-1} while $H_0 = H$ represents the calendar forecast period.
3   Historical Data are denoted by $\{x_{-2,t} \in H_{-2}$ and $x_{-1,t} \in H_{-1}$; t=1, 2,..., n\}; Forecast Values by $\{\hat{x}_t \in H_0$; t=1, ...,k\}.

Algorithm:

1 Cleaning original (raw) data from outliers represents an admissible option.

2 Transformation (1) of historical data to increments of (suitable) reference functions[3] $f_{-2}(t)$ and $f_{-1}(t)$ related to historical data. Transformed data are labeled as P-Index; $P_{-2,t}$ and $P_{-1,t}$; e.g. $P_{-2,t} = x_{-2,t}/f_{-2}(t)$.

3 Transformation (2) of historical data to sequential increments named S-Indices; $S_{-2,t}$ and $S_{-1,t}$; e.g. $S_{-1,t} = x_{-1,t+1}/x_{-1,t}$.

For a 1 period forecast $S_{-1,t} = x_{-1,t+1}/x_{-1,t}$ and $S_{-2,t} = x_{-2,t+1}/x_{-1,t}$; $\forall t$ are calculated.

For a 2 period forecast $S_{-1,t} = x_{-1,t+2}/x_{-1,t}$ and $S_{-2,t} = x_{-2,t+2}/x_{-2,t}$; etc. ...

In general: $S_{-2,t+j} = x_{-2,t+j+1}/x_{-2,t}$ and $S_{-1,t+j} = x_{-1,t+j+1}/x_{-1,t}$ for {j = 1, 2, 3};

4 $P_{0,t} = P_t = a_1 * P_{-2,t} + b_1 * P_{-1,t}$; a_1, b_1 representing Revinda-Coefficients for minimizing forecast errors.

5 Analog step 4 the weighted averages for $S_{0,t+j} = S_{t+j}$ for {j=1, 2, 3} are calculated.

6 $\hat{P}_t = P_t * f_0$ (1) ; $\hat{S}_{t+j} = \hat{P}_t * S_{t+j}$ {j=1, 2, 3}.

7 Final Forecasts $\hat{x}_{t+1}$ represent a linear combination of $\hat{P}_{t+1}$ and $\hat{S}_{t+j}$ for {j=1, 2, 3}.

Parameter specification

Specification of the parameters involved seems to a problem. But that's not true. There is a standard set of initial parameters to be applied for all time series. Those parameters result from experience with many time series. Before using data for forecasting with REVINDA the historical data are analyzed – using a self-made system-tool – including relevant statistical tests. The tool identifies main attributs of the historical time series.

Starting the forecasting process the first forecasts are calculated using the standard parameter values. – In case the forecast accuracy does not meet accuracy targets the parameters will be adjusted according to "optimization criteria". The user is free selecting an accuracy target function. E.g. "Minimize the MAD along the past k forecasting periods". Any other target function is feasible. This procedure represents a WHAT-IF kind of simulation. Based on Newton's Iteration or Gradient-Iteration local minima will be identified resulting in a new set of parameters –comparable to the SOLVER-Function in EXCEL. Experience has shown that target functions including the P_0-Index, which is available for the complete H_0 period, show best results. – Thinking 'down the road' this approach represents a cybernetic cycle adapting the historical (data) process to meeting actual values according to the selected optimization criteria. But the resulting 'new' historical process is not unique. Many solutions are valid. Finding the criteria for selecting the "best" new historical process might challenge research.

Practical experiences with REVINDA

REVINDA enjoys about 3 years application for a rather big retail chain running about 300 outlets in Germany for daily forecasts 2 weeks ahead. As a service for the LSP daily B2C orders and B2B orders from the outlets are forecasted for 24 hours delivery service – given accuracy targets on a daily basis of 75% - 120% of Actuals. The company has established the organization of Business Process Forecast as described in Section 5a. REVINDA is being applied on a weekly basis. - Adjustments of the parameters were needed just twice within 3 years. A visual display is given in Appendix 1 (3).

---

[3] For all periods H_{-2} and H_{-1} and H_0 the selected function-type of f(t) (e.g. linear, polynomial, log, exponential, ...) remains the same. The shift in y-direction for each period depends on the time series data. For some categories of business time series (e.g. Demand, Orders, Sales, ...) $f_0(t)$ can be conjoined with company plans.

5.2 "ERP-Type"- Forecasting - METRIX Approach

The METRIX Approach provides simple forecasting methods based on different similarity metrics. The description in this paper is restricted to just one metric for explaining the idea behind. In practice there are methods for 3 different metrics plus 2 applications on modified data sets (P-Index <see REVINDA-Approach-Algorithm in Section 7.1> and on cumulated data. The final version will provide the "best" Metrix-Forecast (out of 5) given a predefined accuracy measure (e.g. MSE, MAD, MAPE, MASE, ...).

Forecast Options:
The METRIX-Approach offers a k-Step forward forecast up to the end of the calendar forecast period. Data organization will allow for forecasts into the forecast period H_1(H_0 represents the forecast period following historical data.) – METRIX has not been designed for direct daily forecasts.

Data Requirements:
Data describing 2 basic periods (calendar-periodic data, e.g. annual data per month or week, or 2 user specified basic periods, e.g. of 6 or 3 months – but only for continuous data.

Notations:
1. n, t describe the basic period length (e.g. n=12 or 13 for years; n=52 for weeks; while {t=1, 2, ..., n} represents the time axis.
2. Historical periods are denoted by H_{-2} and H_{-1} while $H_0 = H$ represents the calendar forecast period.
3. Historical Data are denoted by $\{x_{-2,t} \in H_{-2}$ and $x_{-1,t} \in H_{-1}$; t=1, 2,..., n}; Forecast Values by $\{\hat{x}_t \in H_0$; t=1, ...,k}.

Algorithm:
1. Metric (absolut): $d_t (H_{-2}; H_{-1})$ = Min $\{x_{-2,t} ; x_{-1,t}\}$ $\forall\ t$ (t = 1, 2, ...,n)
2. $S_{-2} = \sum_t x_{-2,t}$ und $S_{-1} = \sum_t x_{-1,t}$
3. Metric (relative): $D_t = d_t (H_{-2}; H_{-1}) / (S_{-2} + S_{-1})/2$
4. Forecast: $\hat{x}_t = D_t * G_{t-1}$ with

5.
$$G_{t-1} = \begin{cases} UG & \text{for } A < UG & UG = \hat{P} - k_u * \sigma(S_{-2} ; S_{-1}) \\ A = \frac{x_{-2,t}}{x_{-1,t}} * \hat{P} & & \text{with } k_{u,o} = \sim 2 \text{ (e.g.)} \\ OG & \text{for } A > OG & OG = \hat{P} + k_o * \sigma(S_{-2} ; S_{-1}) \end{cases}$$

$\hat{P}$ represents an overall estimate for H_0 allowing for including Plan/Budget. $\hat{P}$ can be adapted in case of significant deviations of Actuals vs Plan-Forecasts during H_0.

In practice UG and OG might be replaced by $\hat{P}$, i.e. $k_{u,o} = 0$ in case of rather stable volumes over H_{-2}, H_{-1}, (H_0). Errors estimating H_0 of below about 20% will have no significant effect on forecast errors.

6   Conclusions:

For weekly and monthly forecasts **REVINDA** and **METRIX** represent innovative and competitive methods to since long time existing standard forecasting methods, being offered/applied in ERP-Systems. The competitive power of both methods still is under review and the results will be published in due course. First accuracy results from applying 5 alternative similarity metrics in line with the METRIX approach can be found in [13]. For daily business forecasts (B2B and B2C) no comparable method to **REVINDA** has been found in literature. Real business experiences of about 3 years confirm the value of REVINDA and might cause potential users for testing the method.

7   Literature:

[1] Spicher, Klaus: "Mensch-Maschine-System"; LOG.Kompass, 10, 2013; S. 36-37

[2] Spicher, Klaus; Fang, Dianjun: "Helix of Logistics"; ICPR 21; 21st International Conference on Production Research; Stuttgart 2011

[3] J. Garland, R. James, E. Bradley: "Model-free Quantification of Time Series Predictability" Physical Review E. Submitted for review Apr. 27, 2014. Accepted on Oct. 14. Preprint available at arXiv:1404.6823 & SFI WP#14-05-014, 2014.

[4] C. Brandt, B. Pombe: Phys. Review Lett. 88, 174102 (2002)

[5] Y. B. Pesin: Russ. Math. Survey 32, 55 (1977)

[6] K. Petersen: "Ergodic Theory", Cambridge University Press, 1989

[7] S. Makridakis et al. (April–June 1982). "The accuracy of extrapolation (time series) methods: results of a forecasting competition" 1 (2). Journal of Forecasting; pp. 111–153

[8] Scott, J. Armstrong & Lusk, Edward, J.: "Commentary on the Makridakis Time Series Competition (M1-Competition)"; *Journal of Forecasting*, 2 (1983), 259-311.

[9] S. Makridakis et al. (April 1993). "The M-2 Competition: a real-time judgmentally based forecasting study" 9. International Journal of Forecasting; pp. 5–22.

[10] S. Makridakis, M. Hibon, (Oct.–Dec. 2000). "The M-3 Competition: results, conclusions And implications"; International Journal of Forecasting; Retrieved April 19, 2014

[11] S. Kolossa: "Can we obtain valid Benchmarks from published surveys of Forecast Accuracy" Foresight, Issue 11, Fall 2008

[12] McCarthy, T.M., Davis, D.F., Golicic, S.L. & Mentzer, J.T.: "The evolution of sales forecasting management: A 20-year longitudinal study of forecasting practice, *Journal of Forecasting*, (2006), 25, 303-324.

[13] Spicher, Klaus: "Neue Prognoseverfahrensansätze und deren praktische Erprobung"; GOR - Vortrag AG Prognoseverfahren; Ingolstadt 8./9. Juni 2015

Appendix 1: REVINDA Examples (Sinus, Random, Daily company data;)

Examples are given for applying the REVINDA-Approach on
1   A "time series" with no random noise (function f(x) = A*sin(a*x +B))
2   A random process (100% random)
3   Company data (scale modified) – for daily forecasts. (The method for calculating daily forecast has not been presented in this paper).

1   Forecasting a periodic function (0% Random Noise)

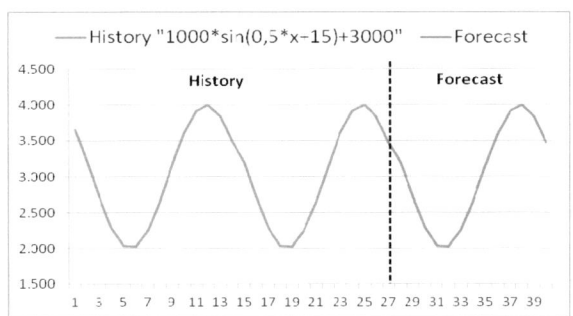

Exhibit 1

The REVINDA-Method forecasts exactly the 'historical' sin-function.

2   Forecasting "100% Random Noise"

Exhibit 2

The forecast represents the same Random Process according to statistical tests.

3 Daily forecasts require organizing the business year according to 13*4 weeks (28 days) data for calendar synchronizing reasons. The data of Exhibit 3 represent modified company data. The weekly structure of the daily profile is based on ERP-type of data processing.

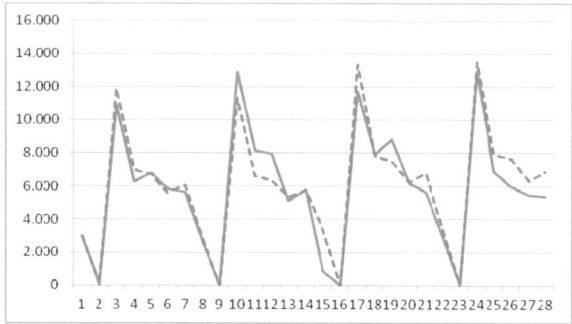

Exhibit 3 Daily Forecasts (dotted) vs Actuals